YOUR KNOWLEDGE HAS VALUE

- We will publish your bachelor's and master's thesis, essays and papers

- Your own eBook and book - sold worldwide in all relevant shops

- Earn money with each sale

Upload your text at www.GRIN.com and publish for free

Bibliographic information published by the German National Library:

The German National Library lists this publication in the National Bibliography; detailed bibliographic data are available on the Internet at http://dnb.dnb.de .

This book is copyright material and must not be copied, reproduced, transferred, distributed, leased, licensed or publicly performed or used in any way except as specifically permitted in writing by the publishers, as allowed under the terms and conditions under which it was purchased or as strictly permitted by applicable copyright law. Any unauthorized distribution or use of this text may be a direct infringement of the author s and publisher s rights and those responsible may be liable in law accordingly.

Imprint:

Copyright © 2017 GRIN Verlag, Open Publishing GmbH
Print and binding: Books on Demand GmbH, Norderstedt Germany
ISBN: 9783668483200

This book at GRIN:

http://www.grin.com/en/e-book/370322/organometallics-fundamentals-and-applications

Kiran V. Mehta

Organometallics: Fundamentals and Applications

GRIN Publishing

GRIN - Your knowledge has value

Since its foundation in 1998, GRIN has specialized in publishing academic texts by students, college teachers and other academics as e-book and printed book. The website www.grin.com is an ideal platform for presenting term papers, final papers, scientific essays, dissertations and specialist books.

Visit us on the internet:

http://www.grin.com/

http://www.facebook.com/grincom

http://www.twitter.com/grin_com

To

My Cheerful Parents: Vasantbhai & Kantaben Mehta

who always give me inspiration, enthusiasm and energy to enjoy life beautifully….!

–Kiran V. Mehta

Preface

Organometallic compounds are extensively explored and used from many years. This book will become a good source of understanding about organometallics.

I hope that this book will be useful to entire scientific community. First two chapters of this book give comprehensive coverage about fundamentals of organometallic compounds. The third chapter describes about the literature available to study this topic. Fourth chapter describes about the general methods for the preparation of organometallic compounds. Fifth chapter of the book is about the research work carried out in the field of organometallics. I hope that my book will serve as a reference source for the persons who are interested in the organometallic chemistry.

This book is also a guide to organometallic literature for various levels of students of organometallic chemistry.

I would like to receive your suggestions about this book.

(Mr.) Kiran V. Mehta

(M.Sc., Ph. D. and Prof. in Chemistry)

Palanpur-385001 (Gujarat) (India)

(2017)

Contents

1.

General introduction

Organometallics

The compounds in which a metal or metalloid is bonded directly to carbon are called organometallics. Here, carbon is present in an organic group. In these compounds, there is a minimum one bond between a metal and carbon atom of an organic compound. In this way, one can say that a metal-carbon bond(R-M bond) is a characteristic of such compounds. In these kinds of compounds, interaction of a metal atom(s) and carbon atom(s) of organic group is important. Compounds of the d-block elements (transition elements), actinide group elements, lanthanide group elements, etc. are these kinds of compounds. Metal is electropositive and carbon is electronegative.

Organometallic have played a foremost role in the progress of Science. They are useful as catalysts and intermediates for the preparation of a large number of compounds.

Organometallic compounds of Lithium and Magnesium are known as Grignard reagents which are also organometallics.

e.g.:

CH_3MgBr
(Methylmagnesium bromide)
CH_3Li
(Methyllithium)
$CH_3 CH_2 CH_2 CH_2Li$
(n-butyllithium)

Triethylborane(Et$_3$B) is an organometallic compound of Boron. Titanocene dichloride, $[Ti(\eta^5-C_5H_5)_2Cl_2]$ is an organometallic compound of Titanium.

In these compounds, there is a σ bond between C atom and the metal. Grignard reagents and organolithiums are excessively alkaline for coupling reactions to prepare alkanes.

Organometallics of Sodium(Na) and Potassium(K) are generally ionics.

Ferrocenes are organometallics in which an iron atom is sandwiched between two hydrocarbon rings.

These compounds are useful for increasing rate of chemical reactions. Compounds like trimethylantimony, trimethylindium, trimethylgallium, trimethylaluminum, etc. are applied in the preparation of semiconductors. For the purification of metals or for the recovery of metals, these compounds are also used. Organometallics can be useful in organic synthesis[1].

History of Organometallics

In 1760, Louis Claude Cadet de Gassicourt found inks based on cobalt salts.

William Christopher Zeise produced Zeise's salt in 1827. It is the first platinum - olefin complex. The formula of Zeise's salt is as below:

$$K[PtCl_3(C_2H_4)] \cdot H_2O$$

Potassium trichloro(ethene)platinate(II).

Zeise's salt is one of the first examples of a transition metal alkene complex. The anion of this complex has η^2-ethylene ligand.

In 1848, Edward Frankland found diethylzinc(DEZ) $(C_2H_5)_2Zn$. He prepared diethylzinc, by heating ethyl iodide in the presence of zinc metal. This is also called Frankland's reagent.

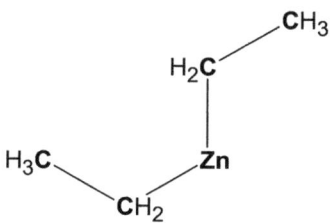

Diethyl zinc

Organochlorosilanes were prepared by Charles Friedel and James Crafts in 1863.

Well known organo nickel complex $Ni(CO)_4$ (tetracarbonylnickel) was found by Ludwig Mond in 1890. It is a toxic substance. In 1899, Grignard complexes were found.

In 1909, arsenic based complex Salvarsan was found by Paul Ehrlich.

In 1951, Walter Hieber was awarded the Alfred Stock prize for his work with metal carbonyl chemistry. Ferrocene has formula: $Fe(\eta^5-C_5H_5)_2$. In 1951, Ferrocene was found.

In 1973, Geoffrey Wilkinson and Ernst Otto Fischer were awarded Nobel Prize for their work on sandwich compounds. In 2005, Yves Chauvin, Robert Grubbs and Richard Schrock were awarded Nobel Prize for their work on metal catalyzed alkene metathesis.

References

[1] Shi L., Cai J. Y. Z., Application of organometallic compounds of groups 15 and 16 in organic synthesis, Highly stereoselective synthesis of Z, E conjugated diene-type sex pheromones, Journal of Organic Chemistry, 1987, 52(16), 3558-3560.

2.

Ways of describing about organometallics and its constituents

Hapticity

F.A. Cotton proposed the term hapticity which was derived from the prefix hapto (from the Greek *haptein*, to fasten denoting contact or combination) placed before the name of the olefin[1].

It is denoted by the Greek letter η (eta) is used to denote the number of contiguous atoms of a ligand which binds to a metal.

Ferrocene and Uranocene can be written as below with this kind of notations:

- Ferrocene

  bis(η^5-cyclopentadienyl) ferrocene-iron

- Uranocene

  bis(η^8-1,3,5,7-cyclooctatetraene)uranium.

- Vasaka's complex

  $IrCl(CO)[P(C_6H_5)_3]_2(\eta^2-O_2)$

η-notation is used when multiple atoms are coordinated. If a ligand coordinates by multiple atoms which are not contiguous, denticity is considered in the place of hapticity.

Denticity

The number of donor groups in a single ligand bound to a metal center atom in a coordination complex is known as denticity.

The word *denticity* is derived from *dentis* (Latin word for tooth). The ligand is considered as biting the metal at one or more linkage points. The denticity of a ligand is denoted by the Greek letter K which is called *kappa*.

If one atom in the ligand is bound to the central metal, denticity is called one. These kinds of ligands are called monodentate ligands or unidentate ligands.

Ligand having more than one bonded atom is known as polydentate ligand. It is also known as multidentate ligand.

EDTA (Ethylenediaminetetraacetic acid) ligand coordinates by six contiguous ligands, so, it is described as K^6 EDTA.

On the basis of denticity, ligands can be classified as below:

(1) Monodentate ligands

These ligands bind to the metal ion with one atom.

(2) Bidentate or didentate ligands

These ligands bind to the metal ion with two atoms.

Ethylene diamine(en) is a bidentate ligand.

(3) Tridentate ligands

These ligands bind to the metal ion with three atoms.

Terpyridine is a tridentate ligand.

These ligands generally bind through two connectivities:

(i) *mer*

It means meridian. In this type of connectivity, the donor atoms are stretched out around one half of the octahedron.

(ii) *fac*

It means facial.

In this type of connectivity, the donor atoms are arranged on a triangle around one face of the octahedron.

1,4,7-triazacyclononane is acyclic ligand.

It is also called as TACN.

It has formula $C_6H_{12}(NH)_3$.

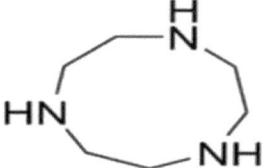

Fig.: Structure of TACN

It binds with fac manner. It is obtained from aziridine(C_2H_4NH).

(4) Tetradentate ligands

These ligands bind with four donor atoms to central metal atom.

Trien (Triethylenetetramine) is a tetra dentate ligand.

(5) Pentadentate ligands

These ligands bind with metal by five donor atoms.

Ethylene diaminetriacetic acid is such a ligand.

(6) Hexadentate ligands

These ligands bind with six donor atoms to central metal atom.

EDTA is this kind of ligand.

(7) Ligands having denticity higher than six

DTPA

It is Diethylene triamine pentaacetate.

DOTA

It is 1,4,7,10-tetraazacyclododecane-1,4,7,10-tetraacetate.

These ligands are useful for lanthanide ions.

References

[1] F. A. Cotton, Proposed nomenclature for olefin-metal and other organometallic complexes, Journal of American Chemical Society, 1968, 90(22): 6230-6232.

3.

Literature about organometallics

Literature of organometallics is of following types:

Primary Literature

Original research papers about organometallics are primary literature. Many reputed journals publish research papers about organometallic compounds. *Journal of Organometallic Chemistry* is one of such journals.

Reviews and Surveys

Origin and development of organometallic compounds are surveyed by many writers. It also gives healthy picture about organometallics. Articles like Personal perspective, inorganic chemistry[1], Concerning organo-magnesium compounds in solution and their application to the synthesis of acids, alcohols and hydrocarbons[2], The discovery of ferrocene, the first sandwich compound[3] and Historical origins of organometallic chemistry Part 1: Zeise's salt[4-5] are good examples of this kind of work.

Textbooks

Various books are available about organometallics. Some of the books are as below:

1. Advanced Practical Inorganic and Metalorganic Chemistry, R. John Errington, CRC Press, 3 Jul 1997.
2. Applications of Organometallic Compounds, Iwao Omae, Wiley, March 1998.
3. Bio-inspired Catalysts (Topics in Organometallic Chemistry), Thomas R. Ward(Editor), Springer, 23 Oct 2014.

4. Comprehensive Organometallic Chemistry II: A Review of the Literature 1982-1994, Cumulative Indexes, Abel Edward F., Stone F., Gordon A.(Editor), Geoffrey W.(Editor), Vol. 14, New York, Pergamon, 1995.

5. Computational Organometallic Chemistry, Olaf Wiest (Editor), Yundong Wu (Editor), Springer, 2012 edition, 29 Feb 2012.

6. Crabtree, Robert H., The Organometallic Chemistry of the Transition Metals, New York, Wiley, 2001.

7. Medicinal Organometallic Chemistry, Edited by Gérard Jaouen (Ecole Nationale Superieure De Chimie De Paris, France) and Nils Metzler-Nolte (Ruhr-Universität Bochum, Germany), Springer, Berlin, Heidelberg, Germany, 2010.

8. Metallocenes, Long Nicholas J., London, Blackwell Science, 1998.

9. Molecular Metal-Metal Bonds: Compounds, Synthesis, Properties, Stephen T. Liddle (Editor), Wiley, March 2015.

10. Occurrence and Analysis of Organometallic Compounds in The Environment, T. R. Crompton, Wiley, March 1998.

11. Organometallic Chemistry, E. G. Rochow, Reinhold, Newyork, 1964.

12. Organometallic Chemistry, Gary O. Spessard, Gary L. Miessler, Oxford University Press, 25 Jun 2015.

13. Organometallic Compounds, G. E. Coates, Methuen, London, 1956.

14. Organometallics, Kiran V. Mehta, Grin Publishing, Germany, 2017.

15. Organometallics in Organic Synthesis, John Melvin Swan, David St. Clair Black, Chapman and Hall distributed in The U.S.A. by Halsted Press, New York, 1974.

16. Organometallics in Process Chemistry, Ahmed F. Abdel-Magid, Springer Science & Business Media, 15 Jun 2004.

17. Organometallics in Synthesis, Fourth Manual, Bruce H. Lipshutz, John Wiley & Sons, 2 Dec 2013.

18. Organometallics in Synthesis, Third Manual, Manfred Schlosser, John Wiley & Sons, 28 May 2013.

19. Organometallics: Complexes with Transition Metal-Carbon [Sigma]-Bonds, Manfred Bochmann, Oxford University Press, 1994.

20. Organometallics: A Concise Introduction, Christoph Elschenbroich, Albrecht Salzer, VCH, 1 Jan 1989.

21. Principles of Organometallic Chemistry, G. E. Coates, Springer, 1st edition, 1 Oct 1968.

22. Schlosser, Manfred (Editor), Organometallics in Synthesis: A Manual, New York: Wiley, 2002.

23. Solid State Organometallic Chemistry: Methods and Applications, Marcel Gielen (Editor), Rudolph Willem (Editor), Bernd Wrackmeyer (Editor), Wiley, October 1999.

24. Spectroscopic Properties of Inorganic and Organometallic Compounds, E. A. V. Ebsworth, Royal Society of Chemistry, 1977.

25. Spessard Gary O., Miessler Gary L., Organometallic Chemistry, Upper Saddle River, NJ: Prentice Hall, 1997.

26. Synthesis and Technique in Inorganic Chemistry: A Laboratory Manual, Gregory S. Girolami, Thomas B. Rauchfuss, Robert J. Angelici, University Science Books, 1999.

27. The Chemistry of Organometallic Compounds, E. G. Rochow, D. T. Hard, R. N. Lewis, Wiley, New York, 1957.

28. The Organometallic Chemistry of N-heterocyclic Carbenes, Han Vinh Huynh, Wiley, February 2017.

29. The Organometallic Chemistry of The Transition Metals, Robert H. Crabtree, Wiley, 6th edition, April 2014.

30. The Organometallic Chemistry of The Transition Metals, Robert H. Crabtree, John Wiley & Sons, 28 Mar 2014.

31. Transition Metal Carbonyl Cluster Chemistry, Paul J. Dyson, J. Scott Mcindoe, CRC Press, November 17, 2000.

References

[1] Cotton F. Albert, A Half-century of nonclassical organometallic chemistry: a personal perspective, Inorganic Chemistry, 2002, 41:643-658.

[2] Jones Paul R., Southwick Everett V. Grignard, Concerning organo-magnesium compounds in solution and their application to the synthesis of acids, alcohols, and hydrocarbons, Journal of Chemical Education, 1970, 47:290-299.

[3] Kauffman, George B., The discovery of ferrocene, the first sandwich compound, Journal of Chemical Education, 1980, 60:185-186.

[4] Thayer John S., Historical origins of organometallic chemistry Part 1: Zeise's Salt, Journal of Chemical Education, 1969, 46:442-443.

[5] Thayer, John S., Historical origins of organometallic chemistry Part 2: Edward Frankland, Journal of Chemical Education, 1969, 46:764-765.

4.

Preparation of

organometallics

Organometallic compounds can give nucleophilic carbons which can react with electrophilic carbons to create a new carbon-carbon bond. Hence, this property of organometallics is useful in the preparation of some complex compounds.

In 1757, Louis Claude Cadet de Gassicourt prepared an organometallic compound which was obtained from arseneous oxide (As_2O_3) and potassium acetate(CH_3COOK).

Salvarsan is another organometallic(organoarsenic) compound. Its activity against syphilis was discovered. It can be prepared as follows:

Edward Frankland prepared the first organozinc compound diethylzinc in 1848 by the reaction between zinc metal and ethyl iodide.

$$2R\text{-}X \quad + \quad 2Zn \quad \longrightarrow \quad R_2Zn \quad + \quad ZnX_2$$

Grignard reagents were prepared by the reaction of an alkyl or aryl halide on magnesium metal.

$$\begin{array}{ccccc} R\text{-}X & + & Mg & \longrightarrow & R\text{-}Mg\text{-}X \\ \text{Alkyl/aryl halide} & & \text{Magnesium} & & \text{Grignard reagent} \end{array}$$

The preparation of the alkyltin and alkyllead was reported by Carl Jacob Lowig in 1852-53.

Around 1917, Wilhelm Johann Schlenk synthesized organolithium compounds by the following route:

$$R\text{-}X \quad + \quad 2Li \quad \longrightarrow \quad R\text{-}Li \quad + \quad LiX$$

Organolithium
compound

He investigated together with his son that organomagnesium halides(RMgX) can participate in a complex chemical equilibrium, which is known as a *Schlenk equilibrium.*

Karl Ziegler directly synthesized lithium alkyls and aryls from metallic lithium and halogenated hydrocarbons.

Ziegler is well-known for his work with Giulio Natta. Their Ziegler-Natta catalysts are based on titanium compounds and organometallic aluminium compounds like, triethylaluminium, $(C_2H_5)_3Al$. These compounds are useful for the polymerization process of 1-alkenes.

$$n\ CH_2 = CHR \quad \longrightarrow \quad [CH_2 - CHR]_n -$$

Ziegler and Giulio Natta won the Nobel Prize in Chemistry in 1963.

Many organometallics can be synthesized from the metal by the oxidative addition of alkyl halides. Halide reactivity increases in the following order:

$$Cl < Br < I$$

The synthesis of organometallics by the reaction between organic substrates with metal powder is a good approach in this direction.

The general methods for the synthesis of organometallics include following reactions:

Method for the preparation of a Grignard reagent is as below:

$$M \quad + \quad RX \quad \longrightarrow \quad RMX$$

Metal vapour can be reacted with a proper substrate to get organometallics:

$$M(vapor) + Substrate \longrightarrow RM$$

Metals reacting with CO give metal carbonyles.

$$M + n\,CO \longrightarrow M(CO)_n$$

Compounds like cyclopentadienyle sodium can be prepared as below:

$$2M + 2RX \longrightarrow 2RM + H_2$$

Hydrogen exchange process can give organometallic:

$$RM + R'H \longrightarrow R'M + RH$$

The reaction of metal halogen exchange process is also used for this kind of preparation:

$$RM + R'X \longrightarrow R'M + RX$$

Sometimes, Insertation reactions are also useful:

$$RM + A \longrightarrow R\text{-}A\text{-}M$$

Decarbonylation can produce organometallic as below:

$$RCOM \longrightarrow RM + CO$$

Synthesis of Acetylenic compounds

The terminal acetylenes can be deprotonated by $NaNH_2$ (sodium amide).

$$RC\equiv CH \xrightarrow{\text{NaNH}_2} RC\equiv CNa + NH_3$$

Acetylenic Grignard reagents, $RC \equiv CMgX$, can also be synthesized.

$$RC\equiv CH \xrightarrow{\text{R'-MgX}} RC\equiv CMgX + R' - H$$

Synthesis of R_2CuLi

Lithium dialkylcuprates, R_2CuLi are organo copper compounds which can be prepared as follows:

$$2RLi + CuX \longrightarrow R_2CuLi + LiX$$

Here, X= Cl, Br, I.

R= alkyl, vinyl or aryl group.

The reactivity of halides is of following order:

I > Br > Cl

Synthesis of RZnX

This can be prepared by oxidation- reduction reaction in the presence of ether:

$$RX \quad + \quad Zn \quad \longrightarrow \quad RZnX$$

These compounds are not so reactive to aldehydes and ketones as alkyl lithium and Grignard reagents are. These organozincs are useful in Simmons-Smith reaction which is shown as below:

$$I-CH_2-I \; + \quad Zn \quad \xrightarrow{\quad Cu \quad} \quad ICH_2ZnI$$

$$ICH_2ZnI \; + \quad H_3C \overset{CH}{\diagdown} \overset{CH_3}{\underset{CH}{\diagup}} \quad \xrightarrow{\quad Et_2O \quad} \quad \triangle \; + \; ZnI_2$$

In this way, Cyclopropanes using organozincs can be synthesized.

5.

Usage of

organometallics

Some organometallic compounds are useful as anticancer agents. In an article, Anna Leonidova and Gilles Gasser reviewed that Rhenium complexes showed anti-proliferative activity and can be used as anticancer drugs[1].

Dean F. Martin reported a necessity for new ions in coordination chemistry. These ions would ease studies of less known metals and of less known coordination numbers and would help studies of chemical bonding and reaction mechanisms[2].

An overview of organometallic: a nonlinear optical (NLO) polymer is presented by Michael E. Wright et al. Organometallic NLO-phores were synthesized by the condensation of ferrocenecarboxaldehyde with fluorene compounds. They exhibited linear optical properties. NLO-spectroscopy was employed for the study the orientation and relaxation behavior of the new organometallic NLO polymers[3].

In an article about the future of organometallic chemistry, Warren E. Piers hoped that organometallic chemistry will be useful to water splitting driven by the sunlight and will constitute a growing area of focus in the coming years[4].

A useful discussion of photo substitution chemistry in organometallic molecules was presented by Gregory L. Geoffroy[5].

Milan Melník et al. reviewed structural parameters of over fifty monomeric organoplatinum complexes with PtP_3C inner coordination sphere. Such complex compounds can be divided into following groups[6]:

$Pt(PL)_3(CL)$

$Pt(PL)_2(\eta^2\text{-}P,CL)$

$Pt(\eta^2\text{-}P_2L)(PL)(CL)$

$Pt(\eta^2\text{-}P_2L)(\eta^2\text{-}P,CL)$

$Pt(\eta^3\text{-}P_3L)(CL)$

Copper complexes containing polyamine chelate ligands are catalysts for ATRP (atom transfer radical polymerization). Zerk T.J. et al. studied such complexes[7].

R. Poli et al. studied the radical polymerization of styrene facilitated by the diaminobis (phenolate) complexes. They indicated that at the higher pressure, the balance between atom transfer radical polymerization (ATRP) and organometallic mediated radical polymerization (OMRP) may shift in the favour of the latter[8].

Iron catalyzed C-C cross couplings using organometallic complexes were reported by Guerinot A. et al. in 2016. This kind of cross-coupling is important for the formation of C-C bonds which may be helpful in the preparation of bioactive molecules[9].

One-Pot synthesis of α-Branched N-Acylamines via Titanium-mediated condensation of Amides, Aldehydes and Organometallics was reported by Dai C. et al. It gives α-branched N-acylamine in high yields[10].

Perylenediimide-isocyanide N,N'-bis(1-ethylpropyl) -1-isocyanide-perylene-3,4:9,10-bis (dicarboximide) (CN-PDI) and isocyanide [AuY(CN-PDI)] and carbene [Au-Y-{C(NH-PDI) (NMe$_2$)}] (Y = Cl, C$_6$F$_5$, 4-C$_6$F$_4$OC$_6$H$_{13}$, 1/2-μ-C$_6$F$_4$C$_6$F$_4$) gold complexes were prepared and characterized by Cristina Domínguez et al. These complexes were coloured. So it can be used as organometallic dyes[11].

Synthesis and characterization of cationic isostructural *'sandwich'* diimine-coinage ethylene complexes of copper aurum were studied by Klimovica K. et al. In this work, ethylene self-exchange kinetics occurs through the associative exchange mechanism for metals[12].

The application of gold complexes in medicine is explored because of their stability. Jurgens S. and Casini A. tested organometallics - Au(I) N-heterocyclic carbenes (NHCs) and cyclometalated Au(III) compounds. It has interesting results[13].

The formation of Gold(III) alkyls from Gold alkoxide complexes is recently reported by Chambrier I. et al.[14].

Synthesis, stability and biological behavior of Thiomaltol (S, O coordinating compound)-based organometallic complexes with 1-methylimidazole as leaving group were studied by Hackl C. M. et al. They prepared Ru^{II}, Os^{II}, Rh^{III} and Ir^{III} complexes having a *'piano-stool'* configuration. They also prepared 1-methylimidazole derivatives. The complexes were analysed by elemental analyses, NMR spectroscopy and X-ray diffraction. ESI mass spectroscopy was used by them for the study in aqueous solution and interactions with certain amino acids. The cytotoxicity against cancer cell lines also investigated by them[15].

Westerhausen et al. prepared heavy Grignard reagents and studied their properties. They prepared aryl-, alkenyl- and alkylcalcium halides. They discussed NMR parameters and ligand redistribution. They also reported reactivity studies with reference to metalation and addition to unsaturated organic compounds and metal-based Lewis acids[16].

Organometallics can be useful in stereo selective organic synthesis of sex pheromones[17].

Alkenes were found to be chelating groups to Zn(II), in stereoselective additions of organozincs to β,γ-unsaturated ketones[18]. Synthesis of piperazines by irradiation and organometallic compounds was reported by Suarez Pantiga S. et al.[19].

Organometallic medicinal chemistry exhibits biomedical and bioanalytical usage of organometallic complexes. Albada B. and Metzler-Nolte N. described biomedical properties of peptides[20].

Uršič M. et al. prepared following new organometallic complexes:

$[(\eta^6\text{-}p\text{-cymene})Ru(4,4,4\text{-trifluoro-1-(4-bromophenyl)-1,3-butanedione})Cl]$

$[(\eta^6\text{-}p\text{-cymene})Ru(4,4,4\text{-trifluoro-1-(4-bromophenyl)-1,3-butanedione})pta]PF_6$

[(η^6-*p*-cymene)Ru(4,4,4-trifluoro-1-(4-iodophenyl)-1,3-butanedione)Cl]

[(η^6-*p*-cymene)Ru(4,4,4-trifluoro-1-(4-iodophenyl)-1,3-butanedione)pta]PF$_6$

These complexes were synthesized and characterized by Ultraviolet–visible (UV-Vis), Infrared (IR), Nuclear magnetic resonance (NMR) and Mass spectroscopy and single-crystal X-ray diffraction[21].

Pop et al. prepared heteroleptic zinc and cadmium complexes of the type [$\{^{Me}_2N^{\wedge}E^{\wedge}O^R_2\}$M-Nu]$_n$ (M = Zn, Cd, E = S, Se, R = CH$_3$, CF$_3$, Nu = N(SiMe$_3$)$_2$, I, Cl, n = 1-2) by treating the alcohol proteo-ligands $\{^{Me}_2N^{\wedge}E^{\wedge}O^R_2\}$H with [M(N(SiMe$_3$)$_2$)$_2$] (M = Zn, Cd) or [XMN(SiMe$_3$)$_2$] (M = Zn, X = Cl, M = Cd, X = I) in an equimolar ratio. They used multinuclear NMR spectroscopy for its characterization. The solid-state structures of complexes were determined by X-ray diffractometer[22].

Zhang et al. fabricated reproducible organometallic-halide-perovskite-based devices. They used fabrication process to create a photodetector which may be useful in optoelectronic devices[23].

Suzuki et al. developed one-pot synthesis of (nitronyl nitroxide)-gold(I)-phosphine (NN-Au-P) complexes by the use of chloro(tetrahydrothiophene)gold(I), phosphine ligands, nitronyl nitroxide radicals and sodium hydroxide[24].

Rossier J. et al. reported the synthesis of new water-soluble vitamin B$_{12}$ prodrugs bearing metal complexes at the β-upper side of the cobalt center. The complexes having the general design {Co-C[triple bond, length as m-dash]C-bpy-M}, where M represents a cytotoxic metal complex, were synthesized and tested for their cytotoxicity against MCF-7 breast cancer cells[25].

Pingyu Zhang and Peter J. Sadler correctly indicated that the design of organometallic complexes for therapeutic and diagnostic applications in cancer and other areas of medicine present new and exciting research opportunities[26].

References

[1] Leonidova A., Gasser G., Underestimated potential of organometallic rhenium complexes as anticancer agents, ACS Chemical Biology, 2014, 9(10), 2180-2193.

[2] Martin D. F., Organometallic-chelate compounds: organometallic ions as central metal ions, Advances in Chemistry, Chapter-35, 1967, **62,** 555-564.

[3] Wright M. E., Cochran B. B., Toplikar E. G., Lackritz H S., Kerney J. T., Inorganic and organometallic polymers II, ACS Symposium Series, 1994, **572,** Chapter 34, 456-471.

[4] Piers W. E., Future trends in organometallic chemistry: organometallic approaches to water splitting, Organometallics, 2011, 30(1), 13-16.

[5] Geoffroy G. L., Organometallic photochemistry, Journal of Chemical Education, 1983, 60(10), 861.

[6] Melník M., Mikuš P., Organophosphines in organoplatinum complexes: structural aspects of PtP_3C derivatives, Journal of Organometallic Chemistry, 2017, **830,** 62-66.

[7] Zerk T. J., Bernhardt P. V., Organo-Copper(II) complexes as products of radical atom transfer, Inorganic Chemistry, Apr 2017, doi: 10.1021/acs.inorgchem.7b00402.

[8] Ploi R., Shaver M. P., Atom transfer radical polymerization (ATRP) and organometallic mediated radical polymerization (OMRP) of styrene mediated by diaminobis(phenolato)iron(II) complexes: a DFT study, Inorganic Chemistry, Jul 2014, 21;53(14), 7580-7590.

[9] Guerinot A., Cosy J., Iron-catalyzed C-C cross-couplings using organometallics, Topics in Current Chemistry, Aug 2016, 374(4), 49.

[10] Dai C., Genovino J., Bechle B. M., Corbett M. S., Huh C. W., Rose C. R., Sun J., Warmus J. S., Blakemore D. C., One-pot synthesis of α-branched n-acylamines via titanium-mediated condensation of amides, aldehydes and organometallics, Organic Letters, Mar 3, 2017, 19(5),1064-1067.

[11] Domínguez C., Baena M. J., Coco S., Espinet P., Perylenecarboxydiimide-gold(I) organometallic dyes, optical properties and Langmuir films, Dyes and Pigments, May 2017, **140,** 375-383.

[12] Klimovica K., Kirschbaum K., Daugulis O., Synthesis and properties of sandwich diimine-coinage metal ethylene complexes, Organometallics, Sep 2016, 12, 35(17), 2938-2943.

[13] Jürgens S., Casini A, Mechanistic insights into gold organometallic compounds and their biomedical applications, Chimia (Aarau), Mar 2017, 71(3), 92-101.

[14] Chambrier I., Roşca D. A., Fernandez-Cestau J., Hughes D. L., Budzelaar P. H. M., Bochmann M., Formation of gold(iii) alkyls from gold alkoxide complexes, Organometallics, 2017 Apr 10, 36(7), 1358-1364.

[15] Hackl C. M., Legina M. S., Pichler V., Schmidlehner M., Roller A, Dömötör O., Enyedy E. A., Jakupec M. A., Kandioller W., Keppler B. K., Thiomaltol-based organometallic complexes with 1-methylimidazole as leaving group: synthesis, stability, and biological behavior, Chemistry, 2016 Nov 21, 22(48),17269-17281.

[16] Westerhausen M., Koch A., Görls H., Krieck S., Heavy Grignard reagents: synthesis, physical and structural properties, chemical behavior, and reactivity, Chemistry, 2017 Jan 31, 23(7),1456-1483.

[17] Lilan Shi, Jianhua Yang Zenwei Cai, Application of organometallic compounds of groups 15 and 16 in organic synthesis, Highly stereoselective synthesis of Z, E conjugated diene-type sex pheromones, Journal of Organic Chemistry, 1987, 52(16), 3558-3560.

[18] Raffier L., Gutierrez O., Stanton G. R., Kozlowski M. C., Walsh P. J., Alkenes as chelating groups in diastereoselective additions of organometallics to ketones, Organometallics, 2014 Oct 13, 33(19),5371-5377.

[19] Suárez-Pantiga S., Colas K., Johansson M. J., Mendoza A., Scalable synthesis of piperazines enabled by visible-light irradiation and aluminum organometallics, Angewandte Chemie International Edition, 2015 Nov 16, 54(47),14094-8.

[20] Albada B., Metzler-Nolte N., Organometallic-peptide bioconjugates: synthetic strategies and medicinal applications, Chemical Reviews, 2016 Oct 12, 116(19), 11797-11839.

[21] Uršič M., Lipec T., Meden A., Turel I., Synthesis and structural evaluation of organo-ruthenium complexes with β-diketonates, Molecules, 2017 Feb 20, 22(2). pii: E326.

[22] Pop A., Bellini C., Şuteu R., Dorcet V., Roisnel T., Carpentier J. F., Silvestru A., Sarazin Y., Cadmium complexes bearing Me2N^E^O- (E = S, Se) organochalcogenoalkoxides and their zinc and mercury analogues, Dalton Transactions, 2017 Feb 21. doi: 10.1039/c7dt00279c.

[23] Zhang N, Sun W, Rodrigues SP, Wang K, Gu Z, Wang S, Cai W, Xiao S, Song Q, Highly reproducible organometallic halide perovskite, micro devices based on top-down lithography, Advanced Materials, 2017 Feb 13. doi: 10.1002/adma.201606205.

[24] Suzuki S., Kira S., Kozaki M., Yamamura M., Hasegawa T., Nabeshima T., Okada K., An efficient synthetic method for organometallic radicals: structures and properties of gold(I)-(nitronyl nitroxide)-2-ide complexes, Dalton Transactions, 2017 Feb 21, 46(8),2653-2659.

[25] Rossier J., Hauser D., Kottelat E., Rothen-Rutishauser B., Zobi F., Organometallic cobalamin anticancer derivatives for targeted prodrug

delivery via transcobalamin-mediated uptake, Dalton Transactions, 2017 Feb 14, 46(7), 2159-2164.

[26] Zhang P., Sadler P.J., Advances in the design of organometallic anticancer complexes, Journal of Organometallic Chemistry, 15 June 2017, **839,** 5-14.

Note for Readers

This book is carefully written and produced. Nevertheless, author, editors and publisher do not warrant the information contained therein to be free of errors. Dear readers are advised to keep in mind that data, statements, procedures, illustrations and other items may inadvertently be inaccurate.

YOUR KNOWLEDGE HAS VALUE

- We will publish your bachelor's and
  master's thesis, essays and papers

- Your own eBook and book -
  sold worldwide in all relevant shops

- Earn money with each sale

Upload your text at www.GRIN.com
and publish for free